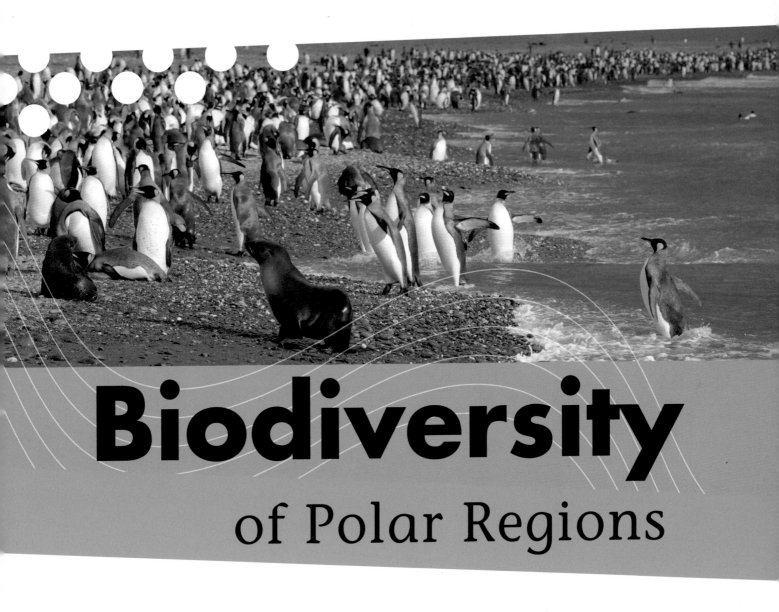

Biodiversity
of Polar Regions

GREG PYERS

MACMILLAN
LIBRARY

First published in 2010 by
MACMILLAN EDUCATION AUSTRALIA PTY LTD
15–19 Claremont Street, South Yarra 3141
Reprinted 2011

Visit our website at www.macmillan.com.au or go directly to www.macmillanlibrary.com.au

Associated companies and representatives throughout the world.

National Library of Australia Cataloguing-in-Publication entry

Pyers, Greg.
   Of polar regions / Greg Pyers.

   ISBN 978 1 4202 6767 9 (hbk.)
   Pyers, Greg. Biodiversity.
   Includes index.
   For primary school age.
   Biodiversity—Polar regions—Juvenile literature. Polar regions—Juvenile literature.

333.950911

Edited by Georgina Garner
Text and cover design by Kerri Wilson
Page layout by Kerri Wilson
Photo research by Legend Images
Illustrations by Richard Morden

Printed in China

Acknowledgements
The author and the publisher are grateful to the following for permission to reproduce copyright material:

Front cover photograph of King penguins and Antarctic fur seals at Salisbury Plain, South Georgia, UK, courtesy of
Photolibrary/Enrique R. Aguirre.
Back cover photograph of a polar bear © yui/Shutterstock.

Photographs courtesy of:
© Carre Claude-BIOSPHOTO/AUSCAPE, 13; © Steven Kazlowski/Science Faction/Corbis, 11, 23; © Paul A. Souders/
Corbis, 28; Paul Chesley/National Geographic/Getty Images, 16; Paul Nicklen/Getty Images, 4; © 2008 Jupiterimages
Corporation, 15; Katy Jensen, National Science Foundation, 24, 27; Patrick Rowe, National Science Foundation, 19;
Photolibrary/Enrique R. Aguirre, 1; Photolibrary/Fred Bruemmer, 26; Photolibrary/David W. Hamilton, 20; Photolibrary/
Geoff Renner, 21; Photolibrary/Robin Smith, 29; Picture Media/REUTERS/HO Old, 17; Picture Media/REUTERS/Icon
Images/Spectral Q/Handout, 25; Shutterstock, 14; © Keith Levit/Shutterstock, 7; © Alistair Scott/Shutterstock, 10; USGS
photograph by Bruce F. Molnia, 22.

While every care has been taken to trace and acknowledge copyright, the publisher tenders their apologies for any
accidental infringement where copyright has proved untraceable. Where the attempt has been unsuccessful, the publisher
welcomes information that would redress the situation.

Please note:
At the time of printing, the Internet addresses appearing in this book were correct. Owing to the dynamic nature of the
Internet, however, we cannot guarantee that all these addresses will remain correct.

Contents

What is biodiversity?   4

Why is biodiversity important?   6

Polar regions of the world   8

Polar biodiversity   10

Polar ecosystems   12

Threats to polar regions   14

    Biodiversity threat: Fishing and whaling   16

    Biodiversity threat: Pollution   18

    Biodiversity threat: Tourism   20

    Biodiversity threat: Climate change   22

Polar region conservation   24

Case study: Antarctica   26

What is the future for polar regions?   30

Glossary   31

Index   32

Glossary words

When a word is printed in **bold**, you can look up its meaning in the Glossary on page 31.

What is biodiversity?

Biodiversity, or biological diversity, describes the variety of living things in a particular place, in a particular **ecosystem** or across the whole Earth.

Measuring biodiversity

The biodiversity of a particular area is measured on three levels:

- **species** diversity, which is the number and variety of species in the area
- genetic diversity, which is the variety of **genes** each species has. Genes determine the characteristics of different living things. A variety of genes within a species enables it to **adapt** to changes in its environment.
- ecosystem diversity, which is the variety of **habitats** in the area. A diverse ecosystem has many habitats within it.

Species diversity

Some types of habitats, such as coral reefs, have very high biodiversity. Others have low biodiversity. The Great Barrier Reef of Australia has around 1600 species of fish, but the entire Southern Ocean has just 120 fish species.

Habitats and ecosystems

There are many habitats in polar regions. Some polar habitats are cliffs, ice shelves, seas, beaches and mossbeds. Different kinds of **organisms** live in these habitats. The animals, plants, other living things and the non-living things, and all the ways they affect each other, make up a polar ecosystem.

The Arctic polar region has diverse species of whales, including narwhals.

Biodiversity under threat

The variety of species on Earth is under threat. There are somewhere between 5 million and 30 million species on Earth. Most of these species are very small and hard to find, so only about 1.75 million of these species have been described and named. These are called known species.

Scientists estimate that as many as 50 species become **extinct** every day. Extinction is a natural process, but human activities have sped up the rate of extinction by up to 1000 times.

Known species of organisms on Earth

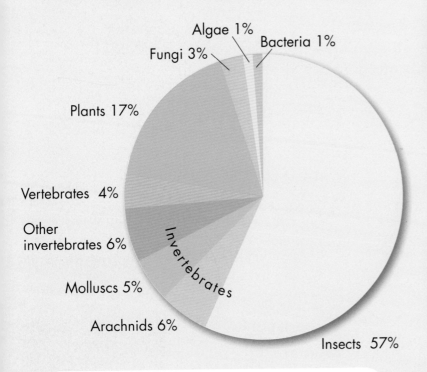

Algae 1%
Bacteria 1%
Fungi 3%
Plants 17%
Vertebrates 4%
Other invertebrates 6%
Invertebrates
Molluscs 5%
Arachnids 6%
Insects 57%

The known species of organisms on Earth can be divided into bacteria, algae, fungi, plant and animal species. Animal species are classified as vertebrates or invertebrates.

Approximate numbers of known vertebrate species

ANIMAL GROUP	KNOWN SPECIES
Fish	31 000
Birds	10 000
Reptiles	8 800
Amphibians	6 500
Mammals	5 500

Why is biodiversity important?

Biodiversity is important for many reasons. The diverse **organisms** in an **ecosystem** take part in natural processes essential to the survival of all living things. Biodiversity produces food and medicine. It is also important to people's quality of life.

Natural processes

Human survival depends on the natural processes that go on in ecosystems. Through natural processes, air and water is cleaned, waste is decomposed, **nutrients** are recycled and disease is kept under control. Natural processes depend on the organisms that live in the soil, on the plants that produce oxygen and absorb **carbon dioxide**, and on the organisms that break down dead plants and animals. When **species** of organisms become **extinct**, natural processes may stop working.

Food

We depend on biodiversity for our food. The world's major food plants are grains, vegetables and fruits. These plants have all been bred from plants in the wild. Wild plants are important sources of **genes** for breeding new disease-resistant crops. If these wild plants were to become extinct, their genes would be lost.

Medicine

About 40 per cent of all prescription drugs come from chemicals that have been extracted from plants. Scientists discover new, useful plant chemicals every year. When plant species become extinct, the chemicals within them are lost forever. The lost chemicals might have been important in the making of new medicines.

Did you know?

Wheat is among the most important of the world's food plants. Its grain is used to make bread, pasta and noodles. Its grass is used as straw to feed animals. Its species name is *Triticum aestivum*. It is a grass, as are other food crops such as rice, rye and barley.

Quality of life

Biodiversity is important to people's quality of life. Animals and plants inspire wonder. They are part of our **heritage**. Some species have become particularly important to us. If the polar bear became extinct, our survival would not be affected, but we would feel great sadness and regret. The polar bear is a powerful symbol of threatened Arctic biodiversity.

Animals such as polar bears inspire people's wonder and imagination. This improves our quality of life.

Extinct species

The great auk was a seabird of the northern Atlantic Ocean. From the 1500s onwards, sailors would load their boats with great auks, as a source of meat.

By 1844, only a few great auks remained, on Eldey Island, off the coast of Iceland. On 3 June 1844, sailors arrived looking for great auks. They found a pair, clubbed them to death and crushed their single egg. The great auk was made extinct.

Polar regions of the world

Polar regions are the areas surrounding the North Pole and South Pole. The Arctic polar region, in the north, is mainly sea and ice. The Antarctic polar region, in the south, is covered almost entirely by the continent of Antarctica.

Boundaries of the polar regions

The world's two polar regions can be defined as the regions lying within the Arctic Circle and within the Antarctic Circle. The Antarctic and Arctic circles are imaginary lines around the Earth. The area north of the Arctic Circle is called the Arctic. The area south of the Antarctic Circle is called the Antarctic.

The Arctic, which is sea ringed by land, can also be defined as the region north of the **tree line**. It can also be defined as the region where the average maximum temperature in the warmest month is 10 degrees Celsius. The edge of this region is marked by the 10 degrees Celsius isotherm.

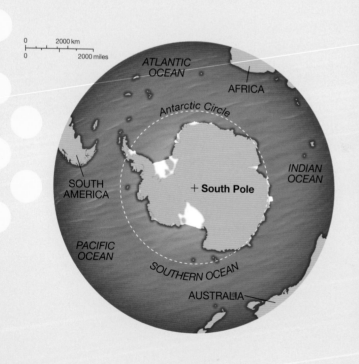

The Antarctic polar region surrounds the South Pole.

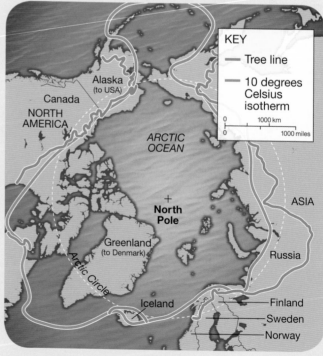

The Arctic polar region surrounds the North Pole.

Polar regions and sunlight

Polar regions are at the northern and southern ends of the Earth's axis, which is the imaginary line around which the Earth rotates. The Earth's axis is tilted at 23.5 degrees. Polar regions are cold areas, because the tilt means that the Sun's rays reach the pole at an angle, so the Sun's rays spread over a larger area and have less intensity.

As the Earth travels around the Sun throughout the year, the poles receive varying periods of sunlight each day. When a pole faces the Sun, it has long periods of sunlight. When it faces away from the Sun, it has long periods of darkness.

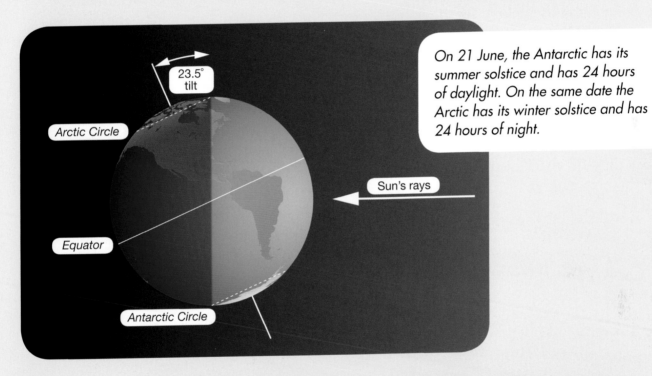

23.5° tilt

Arctic Circle

Equator

Sun's rays

Antarctic Circle

On 21 June, the Antarctic has its summer solstice and has 24 hours of daylight. On the same date the Arctic has its winter solstice and has 24 hours of night.

Summer and winter solstices

If you stood anywhere within the Antarctic Circle at the summer solstice, which is around 21 December each year, you would see that the Sun does not set. There are 24 hours of sunlight. At the winter solstice, around 21 June, the Sun does not rise and there are 24 hours of darkness. Standing within the Arctic Circle, you would make the opposite observations on the same dates.

Polar biodiversity

The biodiversity of the Arctic is quite different from that of the Antarctic. This is mainly because the regions **evolved** separately and are far apart. All polar animals and plants live in harsh conditions, however, so there are some similarities.

Similarities in Arctic and Antarctic biodiversity

Arctic and Antarctic **habitats** are similar, so they support similar kinds of **organisms**. These polar habitats include **ice floes**, rocky **outcrops**, beaches, seas, cliffs, mountains and moss beds.

Plant biodiversity

There are no trees or bushes in polar regions. Rainfall and temperatures are too low, soils are poor and for much of the year there is insufficient light. Only simple plants, such as mosses, lichens and liverworts, survive.

Animal biodiversity

Birds and mammals are the only land-based vertebrates in polar regions. These animals generate their own body heat. They have bodies that are insulated with thick fur or feathers to retain body heat, and some have a layer of **blubber** beneath their skin. Reptiles and amphibians depend on the Sun's heat to warm their bodies and there is too little solar radiation in polar regions for these **species** to survive.

Did you know?

Only a few species are found in both polar regions. Humpback whales are found in Arctic and Antarctic seas, and Arctic terns fly from the Arctic to Antarctica and back every year.

The Arctic tern is found in both Antarctica and the Arctic.

Differences in Arctic and Antarctic biodiversity

Antarctica has been isolated from other landmasses for 30 million years. In this time, many species have evolved and most of these are **endemic species**, which are found nowhere else. The Arctic is not isolated from landmasses, so few of its species are endemic to the region. Plant and invertebrate species are constantly invading from the south.

There are no completely land-based vertebrates in Antarctica. Most Antarctic vertebrates depend on the sea for their food, so they live around the coast. Land-based vertebrates, such as caribou, migrate to the Arctic from the south every summer. Arctic **marine** biodiversity is not as rich as Antarctic marine biodiversity.

In summer, when the Arctic snow melts, caribou move north to feed on exposed plants.

Polar vertebrate diversity

ANIMAL GROUP	ARCTIC SPECIES	ANTARCTIC SPECIES
Land-based carnivores (meat eaters)	Arctic wolf, Arctic fox, polar bear	None
Herbivores (plant eaters)	Musk ox, caribou	None
Seals, sea lions and walruses	Ringed seal, walrus, hooded seal, common seal	Leopard seal, Weddell seal, crabeater seal
Whales	Humpback whale, killer whale, bowhead whale, narwhal, beluga	Humpback whale, killer whale, southern right whale
Flightless birds	None	Seven species of penguins
Birds of prey	Snowy owl	None

Polar ecosystems

Living and non-living things, and the **interactions** between them, make up polar **ecosystems**. Living things are plants and animals. Non-living things are snow, ice and sea water, as well as the **climate**, temperature and ocean currents.

Food chains and food webs

A very important way that different **species** interact is by eating or consuming other species. This transfers energy and **nutrients** from one **organism** to another. A food chain illustrates this flow of energy, by showing what eats what. Food chains are best set out in a diagram. A food web shows how different food chains fit together.

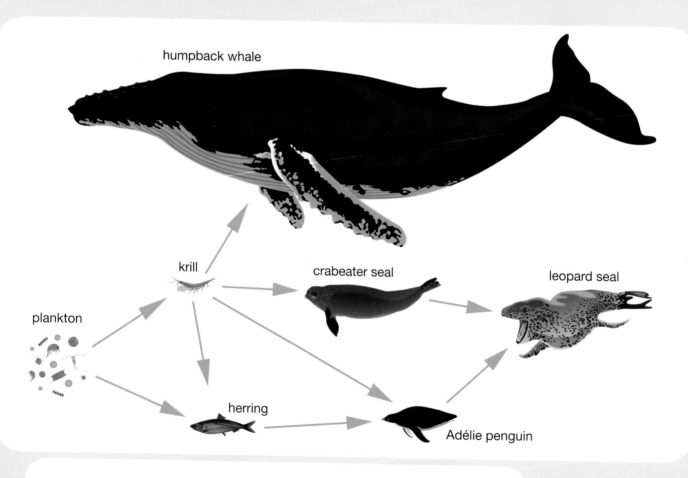

This Antarctic food web is made up of several food chains. In one food chain, plankton is eaten by herrings, which are eaten by Adélie penguins, which in turn are eaten by leopard seals.

Other interactions

Living things and non-living things in a polar ecosystem interact in other ways, too.

Parasitic relationships

Polar birds and mammals have lice. These insects are **parasites** and live on their hosts' bodies. Depending on the species of lice, they may feed on dead skin, feathers or blood.

Day length and behaviour

The behaviour and life cycles of polar animals are affected by the changing day length. At the beginning of winter, male and female emperor penguins walk up to 100 kilometres inland from the sea ice to their breeding sites. The birds sense the shortening day length and know it is time to make the journey.

Temperature and growth

In Antarctica, more sea ice forms in colder winters. More sea ice supports the growth of large species of **phytoplankton**, called diatoms, which are eaten by **krill**. The krill is then eaten by whales and other animals. When winters are not so cold, less sea ice forms and smaller phytoplankton species grow. Krill do not eat these species as much. With less krill, there is less food for whales and other krill-eaters.

Diatoms are microscopic plants that drift in the sea. They play an important part in many Antarctic ecosystems.

BIODIVERSITY THREAT:
Threats to polar regions

The biodiversity of polar regions faces several major threats, such as fishing, whaling, pollution, tourism and **climate** change. More people are moving into polar regions and as polar **icecaps** shrink, increased shipping and mining threaten biodiversity, too.

More tourists and settlements

Increased numbers of humans living in and visiting polar regions threatens polar diversity. Human settlements normally produce pollution, such as oil leaks from machinery and discarded plastics and chemicals. People bring animals and plants with them, too, either accidentally or deliberately. These new **species** can become pests in polar regions.

Shrinking icecaps

Polar icecaps expand in winter and recede in summer. Currently, the icecaps are smaller than they have been for thousands of years. This is because the world's average temperature is increasing. Some scientists believe the Artic's summer icecap may disappear completely after 2030. This would have major consequences for polar biodiversity. Many species may become **extinct**. Other species may invade and take over the new, warmer environment.

Icecap size

The Arctic icecap is frozen sea, called pack ice. It covers an area of between 9 and 12 million square kilometres. The Antarctic icecap covers almost the entire continent, which is 14 million square kilometres of land. In winter, the sea around Antarctica freezes, doubling the area of the icecap.

Chunks of floating ice that break off polar icecaps are called icebergs.

As the Arctic icecap melts, more ships can navigate their way through the Arctic seas.

New shipping route

The melting of the Arctic icecap is likely to open up a new shipping route through the Arctic. Ships will be able to sail from Europe to Asia across the north of Canada. This will increase the risk of oil spills in the Arctic. It will also increase the likelihood that **invasive species** will reach the Arctic, carried in ships' **ballast** tanks.

Gas and oil exploration

Undersea exploration for oil and gas is increasing in the Arctic. Leakages of oil from rigs and underwater pipelines can destroy large areas of seabed. Ninety per cent of the 5000 known invertebrate species in Arctic waters are bottom-dwelling species that live on the seabed. Oil leaks would be catastrophic for these species.

Invasive crabs

Red king crabs are native to Alaskan seas. In the 1960s, they were released off the northern coast of Russia to create a crab-fishing industry for the local people. The crabs thrived but proved to be a major **marine** pest, spreading as far west as Norway by 2004. Red king crabs eat other bottom-dwelling animals so efficiently that in some areas they are the only animals left across whole areas of seabed.

BIODIVERSITY THREAT:
Fishing and whaling

The world's traditional fishing areas are becoming overfished and fishing boats are now harvesting the polar seas. **Krill** fishing, whaling and illegal fishing threaten polar **ecosystems**.

Krill fishing

Krill are the main food for many polar **species**. Humpback, southern right, bowhead and blue whales eat tonnes of krill a day. Crabeater seals, albatrosses and penguins also depend on krill as a large part of their diets.

People have been catching krill for about 30 years. Krill are fished to make fishmeal, which is used as an animal feed, and various health products. The animals are often sucked from the sea using huge tubes. Scientists estimate there are about 100 million tonnes of krill in the Antarctic. Under the rules of the international Convention for the Conservation of Antarctic Marine Living Resources, catches of krill are limited to 4 million tonnes each year. Newer, larger boats are being built, however, and there will be pressure from the fishing industry to increase this figure.

Krill is fished with large nets in the Pacific Ocean. Overfishing of krill threatens the biodiversity of an area.

Whaling

In the 1800s and 1900s, hundreds of thousands of whales were killed in polar seas. Commercial whaling ended in 1986, but many species have yet to build their numbers again. Today, Japan catches around 1000 minke whales a year and wants to catch fin and humpback whales, too. It argues that its whaling is for scientific research and is not a threat to any whale species. Many member nations of the International Whaling Commission, which sets the rules for whaling, disagree with Japan's whaling.

Illegal fishing

Illegal fishing is a threat to the toothfish of the Southern Ocean. Based on scientific studies, limits are set as to how many toothfish can be caught without threatening the species. Illegal fishers ignore these limits. They also ignore international rules on longline fishing, which is the method used to catch toothfish.

A Japanese whaling vessel hauls in a harpooned whale.

Longline fishing

Longline fishing is a method of fishing in which baited hooks are trailed on a line that stretches behind a boat for up to 100 kilometres. Before the hooks sink, seabirds often dive for the bait, get hooked and drown. Every year, around 300 000 seabirds are killed by longline fishing. Twenty of the world's 24 albatross species are threatened, mainly because of longline fishing.

BIODIVERSITY THREAT:
Pollution

Polar regions are a long way from large cities, but pollution is still a major threat to polar biodiversity. Some **pollutants** are transported naturally to polar regions. Oil spills and nuclear waste are also threats.

Pollution in the polar regions

Pollutants are transported towards the polar regions in a process called global distillation. Global distillation is driven by evaporation in warm areas of the world and condensation in cold areas. The pollutants that gather in polar regions through global distillation are called persistent organic pollutants or POPs.

Persistent organic pollutants

Persistent organic pollutants take a long time to break down and some are very toxic chemicals. They damage animals' immune systems, affect their reproduction, and cause cancer, birth defects and death.

POPs are taken in by **organisms** and become concentrated in the body fat of animals at the top of the food chain. This is called bioaccumulation. This process is particularly harmful to polar **predators** such as whales, seals and polar bears. These animals are at the top of their food chains, have large amounts of fat and live a long time, meaning they can accumulate very high levels of POPs.

Bioaccumulation of persistent organic pollutants (POPs)

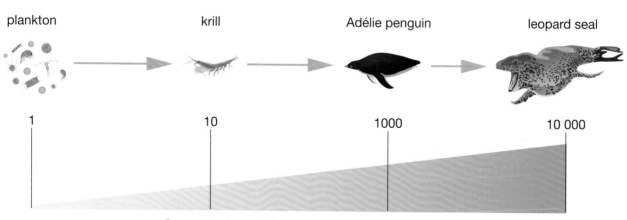

plankton	krill	Adélie penguin	leopard seal
1	10	1000	10 000

Concentration of POPs (milligrams per kilogram)

Adélie penguin colonies in Antarctica are at risk from persistent organic pollutants.

Did you know?

Adélie penguins that spent their entire lives in Antarctica were found to have concentrations of POPs in their bodies up to 100 times higher than those of penguins further north.

Oil drilling

The world's oil reserves are running out and oil is highly valued. Large reserves of oil have been discovered in the polar region of Russia. As sea ice melts, these reserves become easier and less expensive to reach. Because polar animals tend to gather in large numbers, a single oil spill could kill a whole population of a **species**.

Nuclear waste

Radiation is leaking from nuclear waste stored on the Kola Peninsula on the northern coast of Russia. This nuclear waste comes from nuclear-powered ice-breaker ships and submarines. The area's seafloor is so badly contaminated that no fish can live there. The cost of removing the waste is likely to be more than US$4 billion.

BIODIVERSITY THREAT:
Tourism

Today, thousands of people cross the Antarctic Circle into the polar region as tourists. Tourism to the Arctic is also increasing.

Visitors to polar regions

Tourists visit Antarctica between October and April every year. Most come by boat. In 1985, just a few thousand people visited Antarctica but in the 2007–08 season, the number of tourists was more than 40 000. Most tourist trips to Antarctica are to the Antarctic Peninsula where the sea is free of ice for longer than elsewhere on the continent. Some ships carry more than 3000 people each journey. This is more than the total number of people who live at Antarctic research stations.

Up to 1 million tourists visit the Arctic each year. Tourists arrive by boat, plane and land vehicles, such as caterpillar tractors and four- wheel drives. In Russia, the use of ice- breaking ships means that tourists can visit the Arctic even when the sea is frozen.

Did you know?

No human had crossed the Antarctic Circle until 1773, when Captain James Cook became the first person to do so.

A group of tourists gather to photograph penguins in Antarctica. Tourism to Antarctica is increasing.

How can tourism affect biodiversity?

Tourism can threaten biodiversity in several ways. These threats include:

- increased risk of pollution, such as oil spills and plastic waste
- disturbance of breeding animals, which waste energy reserves as they avoid tourist boats and helicopters
- increased risk of introducing **invasive species**
- trampling of **vegetation**
- **erosion** of land caused by vehicle traffic in the Arctic.

Tourists flock to see penguins, polar bears, walruses and seals. These animals tend to gather in small areas and tourists gather in their greatest numbers in these places, too. This means the most important sites for wildlife are put under the most pressure from tourism.

Every footstep that visitors take can disturb polar wildlife. In 2009, the 28 nations of the Antarctic Treaty agreed to limit the size and number of tourist ships visiting the Antarctic.

The importance of management

Many of the threats tourism poses to biodiversity are due to poor management. In the Russian Arctic, trampled vegetation could be avoided by the construction of raised wooden walkways. Disturbance to wildlife could be reduced by not allowing tourists to get too close to nesting birds or basking seals.

BIODIVERSITY THREAT:
Climate change

The world's average temperature is rising because levels of certain gases, including **carbon dioxide**, are increasing in the Earth's atmosphere. This is causing changes to the world's **climate**. These changes are affecting polar biodiversity.

Changing habitats

As the Earth's atmosphere becomes warmer, **habitats** in polar regions are changing. In the Arctic, animal and plant **species** move in from the south and grow on land that was once covered in ice. These species may be carried into polar regions on machinery, so any growth in human activity in polar regions increases the rate at which these changes occur.

In Antarctica, the ice shelf and coastal **glaciers** along the Antarctic Peninsula are shrinking. This is increasing the area of shallow water along the coast. The shallow water warms up easily and the melting ice makes the sea water less salty. These changes affect biodiversity because species that are **adapted** to cold, salty water may not be able to survive in changed conditions. Other species that are adapted to the conditions may take their place.

Coastal glaciers in Alaska, in the Arctic, are gradually shrinking and changing the habitats around them.

Seabirds dive for Arctic cod, which feed on organisms beneath the polar sea ice. Any changes to the sea ice affect this food chain.

The importance of ice to polar biodiversity

Many polar **organisms** are dependent on sea ice for their survival. Seals and penguins use **ice floes** as platforms for resting and polar bears use them for moving across the sea to hunt. Many other species depend on ice, too. The underside of sea ice is habitat for many species of **phytoplankton**, such as diatoms. These tiny organisms make up half the energy-producing plant life in some polar areas. They are at the bottom of the food chain and larvae of various insects feed on them. In turn, the larvae attract fish such as Arctic cod, which are preyed on by seals.

Effects of carbon dioxide on krill

Increased levels of carbon dioxide in sea water may result in a reduction in **krill**. Krill kept in tanks exposed to high levels of carbon dioxide produced deformed larvae or eggs that failed to hatch. Carbon dioxide is absorbed by sea water and makes the sea acidic. This interferes with the krill's ability to make the calcium carbonate in their skeletons. A reduction in krill would be catastrophic for the many species that feed on them.

Polar region conservation

Conservation is the protection, preservation and wise use of resources. Scientists from many countries work on research projects to find out about polar biodiversity, how human activities are affecting it and how it can be conserved.

Research and management

Research surveys or studies are used to find out information about polar **ecosystems**, such as how they work and how humans affect them. Research helps people work out ways of conserving polar ecosystems and how to manage polar areas. The Commission for the Conservation of Antarctic Marine Living Resources is the international organisation responsible for managing fishing and conservation of **marine** wildlife in the Antarctic. In the Arctic, this work is carried out by the Arctic Council, which is made up of the eight countries that have land or sea territory within the Arctic Circle.

A research scientist measures a giant petrel chick in Antarctica.

Did you know?

A 2008 study by WWF found that up to 75 per cent of Antarctica's penguin **species** are likely to decline drastically or become **extinct** due to **climate** change.

Education

Educating people about polar conservation is very important. When people know about a place, they are more likely to want to care for it. The best way to find out about polar biodiversity is to visit a polar region as a worker at a research station or as a tourist. Despite the threats posed by tourism, this industry can be very beneficial to polar conservation. The director of the International Association of Antarctica Tour Operators believes that tourists become ambassadors for polar conservation. This means they return home and tell their friends about their experience of the beauty of the polar environment and how it should be protected.

World Parks Congress

The World Parks Congress is run by the IUCN, the International Union for the Conservation of Nature, and takes place every ten years. The congress makes recommendations for biologically important areas to be protected as reserves and national parks. At the last Congress, in 2003, the Ross Sea of Antarctica was highlighted as a high priority for protection as 'the largest relatively intact marine ecosystem remaining on Earth'.

Anti-whaling protesters try to raise awareness of whaling and conservation issues.

CASE STUDY:
Antarctica

Most Antarctic life is **marine** life, found around the coast. Both its marine and land-based **organisms** need to be protected from the negative effects of human settlement and tourism.

Antarctic marine biodiversity

Compared with other seas and coastlines of the world, the Antarctic has very high marine biodiversity and many new **species** are discovered every year. Antarctic waters are rich in invertebrate species. They have a very high proportion of the world's species of sea spiders but a low number of barnacle, crab, shark and ray species. Sea jellies and amphipods, a type of crustacean, are in great variety in Antarctic waters.

The seabed around Antarctica is covered in animals such as sea stars, brittle stars, sea cucumbers, sponges, anemones, snails and lamp shells. **Phytoplankton** are in large quantities.

Antarctic land-based biodiversity

There are more than 200 species of lichens and 50 species of mosses in Antarctica. There are only two species of flowering plants, the Antarctic hair grass and the Antarctic pearlwort, which are both found only on the Antarctic Peninsula. Animals that live entirely on land in Antarctica are nearly always invertebrates, such as springtails, lice and mites.

King penguins and fur seals are part of Antarctica's marine and coastal biodiversity.

The rich biodiversity of the Ross Sea

The Ross Sea is a marine **ecosystem** that is almost completely undisturbed by humans. It is home to the greatest biodiversity of all the polar regions. It surrounds less than 10 per cent of the Antarctic coast and lies over a continental shelf area as large as Europe.

Ross Sea biodiversity

SPECIES	COMPARISON
Adélie penguin	38% of world's population
Emperor penguin	26% of world's population
Antarctic petrel	30% of world's population
Minke whale	21% of Antarctic population
Killer whale (dwarf form)	More than 50% of world's population
Weddell seal	50% of population of Pacific part of Antarctica

Threats to Ross Sea biodiversity

Recently, commercial fishing has begun in the Ross Sea. The target species is the Antarctic toothfish. This species is eaten by seals and penguins and also competes with them for fish. The commercial fishing of toothfish will have an effect on the numbers of toothfish, seals and penguins but it is not known to what level it will affect them.

Antarctic toothfish are fished in the Ross Sea, off the coast of Antarctica.

27

Protecting Antarctica's biodiversity

Antarctica's biodiversity needs to be protected from pollution, such as waste from research stations, and the negative effects of tourism.

Disposing of waste

Around 50 research stations have operated in Antarctica over the past 100 years. Until the mid-1980s, much of the waste produced by these stations was dumped in rubbish tips nearby. Old oil drums and discarded batteries leaked onto the land and plastic rubbish was blown across the surrounding landscape. Burning rubbish was common, as was bulldozing rubbish onto **ice floes** in winter so it would be carried off when the ice broke up in summer. Sewage was often pumped into the sea. This waste was unsightly, but it also threatened wildlife. By 1990, there was an estimated 1 million cubic metres of rubbish in Antarctica.

Did you know?

In 2004, the Australian Antarctic station Wilkes had approximately 30 000 tonnes of waste. This may take up to 10 years to remove.

In 1991, countries with research stations in Antarctica agreed that waste disposal practices had to change. They signed the Protocol on Environmental Protection to the Antarctic Treaty, which declared that Antarctica was a natural reserve 'devoted to peace and science'. The Protocol states that human activities should be planned to limit adverse impacts on the Antarctic environment.

Rusted, discarded equipment and waste are left behind in Antarctica.

Managing tourism

Tourism to Antarctica is a growing industry. Tourist operators, who organise and conduct tourist trips, understand that their industry depends on protecting the wildlife that people come to see. In 1991, tourist operators set up the International Association of Antarctica Tour Operators (IAATO). This organisation has 100 members from 15 countries. The IAATO writes guidelines and rules that ensure that Antarctic **habitats** and wildlife are not harmed by inappropriate tourist activities. These guidelines and rules control the number of tourists allowed per visit, how closely tourists can approach animals and how rubbish is disposed. It even set a rule that the soles of footwear should be clean, in case seeds from other countries are carried in mud stuck in a shoe's tread. These seeds could sprout and become weeds.

Huskies

The Protocol on Environmental Protection to the Antarctic Treaty 1991 bans the taking of any non-native animal **species** to Antarctica. This meant that husky dogs, which had been used in Antarctica as sled dogs since 1898, had to be removed. In 1993, the last remaining huskies at Mawson, an Australian base, were taken away to live out their days in the United States.

Tourists watch a crabeater seal from a distance. There are laws about how close tourists may get to wildlife.

What is the future for polar regions?

Pollution, tourism, fishing and whaling can be controlled if people decide that polar biodiversity is important enough to protect. Whaling can be banned and Antarctic research stations can be cleared of rubbish. **Climate** change remains the greatest threat.

What can you do for polar regions?

You can help protect polar regions in several ways.

- Find out about polar regions. Why are they important and what threatens them?
- Become a responsible consumer. Do not litter, reduce your use of packaging and walk rather than take the car.
- If you are concerned about polar regions, write to or email your local newspaper, your local member of parliament or another politician and tell them your concerns. Know what you want to say, set out your argument, be sure of your facts and ask for a reply.

Useful websites

💻 **http://wwf.panda.org/what_we_do/where_we_work/arctic/**
This WWF website has information about the Arctic polar region and conservation work in this area.

💻 **http://www.biodiversityhotspots.org**
This website has information about the richest and most threatened areas of biodiversity on Earth.

💻 **http://www.iucnredlist.org**
The IUCN Red List has information about threatened plant and animal species.

Glossary

adapt   change in order to survive

ballast   sea water that is taken on board a ship to keep it weighted and stable at sea when it is not carrying cargo

blubber   the outer layer of fat of sea mammals, such as seals and whales

carbon dioxide   a colourless and odourless gas produced by plants, animals and the burning of coal and oil

climate   the weather conditions in a certain region over a long period of time

ecosystem   the living and non-living things in a certain area and the interactions between them

endemic species   species found only in a particular area

erosion   wearing away of soil and rock by wind or water

evolved   changed over time

extinct   having no living members

genes   segments of deoxyribonucleic acid (DNA) in the cells of a living thing, which determine characteristics

glaciers   rivers of ice that flow very slowly down mountains

habitats   places where animals, plants or other living things live

heritage   things we inherit and pass on to following generations

icecaps   layers of ice over large areas, particularly the poles

ice floes   sheets of floating ice

interactions   actions that are taken together or that affect each other

invasive species   non-native species that spread through habitats

krill   small shrimp-like animals that live in huge swarms

marine   of the sea

nutrients   chemicals that are used by living things for growth

organisms   animals, plants and other living things

outcrops   rock formations that are visible above the ground

parasites   organisms that live on or in organisms of another species

pollutants   harmful or poisonous human-produced substances that enter an environment, possibly causing damage to organisms

phytoplankton   microscopic plants that drift in the sea

predators   animals that kill and eat other animals

species   a group of animals, plants or other living things that share the same characteristics and can breed with one another

tree line   an imaginary line beyond which is it is too cold for trees to grow

vegetation   plants

Index

A

albatrosses 16, 17
Antarctica 8, 10, 11, 13, 14, 19, 20, 22, 24, 25, 26–8
Antarctic research stations 20, 25, 28, 30
Arctic region 8, 9, 10, 11, 14, 15, 20, 21, 22, 24

B

biodiversity hotspots 30

birds 5, 7, 10, 11, 13, 17, 21

C

climate change 14, 22–3, 24, 30
conservation 16, 24–5
crabs 15, 26

E

ecosystem diversity 4, 6–7, 10, 11, 12–13
ecosystems 4, 6, 12–13, 16, 24, 25, 27
education 25
endemic species 11
extinct species 5, 6, 7, 14, 24

F

fish 4, 5, 17, 19, 23, 27
fishing 14, 15, 16, 17, 24, 27, 30
food chains 12, 13, 23
food webs 12

G

genetic diversity 4, 6
great auk 7

H

habitats 4, 10, 22, 23, 29
husky dogs 29

I

icecaps 14
ice floes 10, 23, 23

K

krill 13, 16, 23

L

longline fishing 17

M

medicines 6, 7
microhabitats 4

N

North Pole 8
nuclear waste 18, 19
nutrients 6

O

oil drilling 15, 19

P

parasites 13
penguins 11, 13, 16, 19, 21, 23, 24, 27
persistent organic pollutants 18
polar bears 7, 11, 18, 21, 23
pollution 14, 18–19, 21, 28, 30

R

research 17, 24, 25
Ross Sea 25, 27

S

seals 11, 16, 18, 21, 23, 27
shipping 14, 15
South Pole 8
species diversity 4, 5, 10, 11, 26, 27
sunlight hours 9

T

temperature increases 14, 22
threats to biodiversity 5, 14–15, 16–17, 18–19, 20–21, 22–3, 25, 27, 28 30
toothfish 17, 27
tourism 14, 20–21, 25, 26, 28, 29, 30
tree line 8

W

waste disposal 19, 28
websites 30
whales 10, 11, 13, 16, 17, 18, 27
whaling 14, 16, 17, 30